The Tiger
of the River

Written by
Adrian Pinder

Illustrated by
Maya Ramaswamy

talking
CUB
An Imprint of Speaking Tiger Books

TALKING CUB
Published by Speaking Tiger Books LLP
125A, Ground Floor, Shahpur Jat
New Delhi 110049

First published in paperback in Talking Cub by Speaking Tiger Books in 2022
Text copyright © Adrian Pinder 2022
Illustrations copyright © Mahseer Trust 2022

ISBN: 978-93-5447-171-1
eISBN: 978-93-5447-179-7

10 9 8 7 6 5 4 3 2 1

Printed by : SBM Industries Pvt Ltd, Rai (www.sbmprints.com)

For Jasmine, Harry and Rudi

And in loving memory of Clive Pinder

All life on Earth
needs water
to survive
and grow.

Without water, there would be no trees, no animals and no people.

We humans worship rivers, but sometimes we forget how precious they are.

This is a story about a mighty river called the Kaveri which flows across South India—and all the creatures, big and small, who live in and around it.

It is difficult for humans to see beneath the water, but in the river, lives a very special fish called the mahseer.

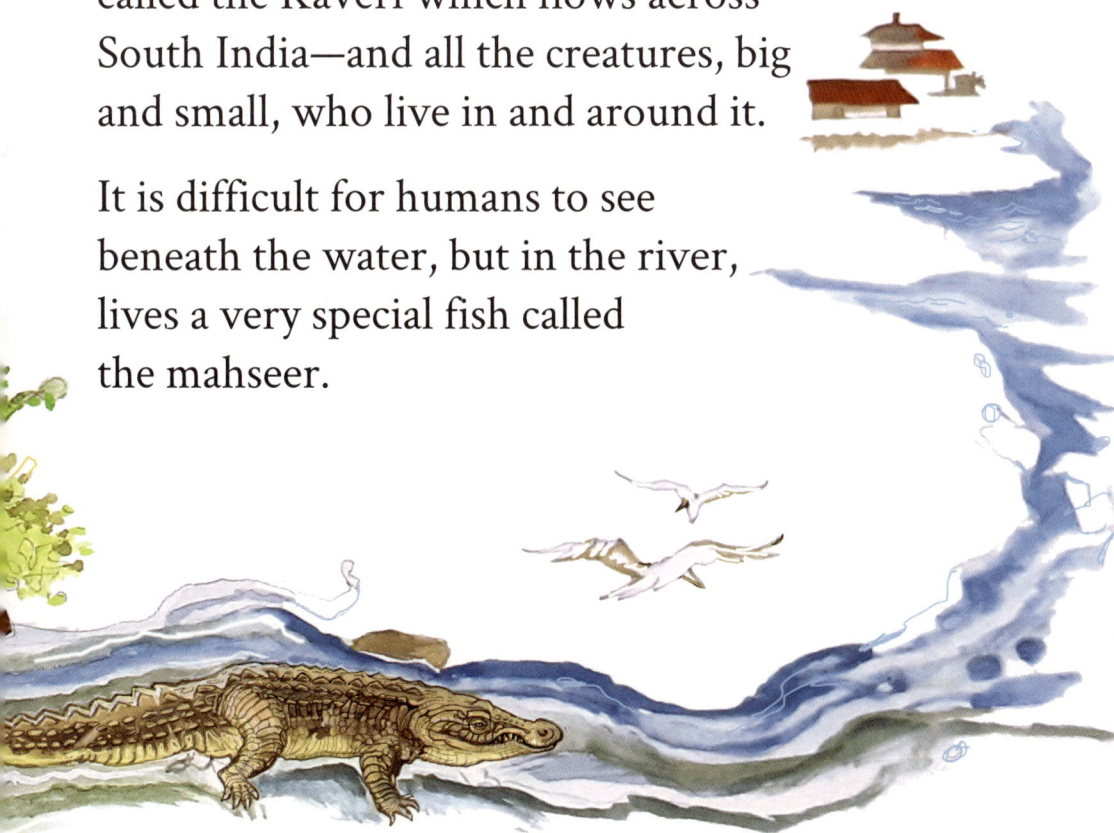

As the bright sun slowly rose in the sky, an orange glow was cast on the Kaveri River valley. The sun's heat burned away the morning mist which was clinging to the river. Slowly, the mist swirled across the rocky gorge of Mekedattu.

On the cliff, a troop of langur monkeys were playing. Beneath, the river roared. It was full of water from the recent monsoon rains.

As the water raced past the playing monkeys, it carried many interesting items with it.

Amongst the fallen trees and plastic litter left by humans, a tiny fish (*no larger than a grain of rice!*), was caught up in the flow and helplessly washed downstream.

The river carried Matisha, the little fish, to the safety
of a sheltered bay, along with many other baby fish.

Matisha did not know she was different from the other little fish in her shoal. She played happily with them in the warm water where there were lots of delicious little bugs for them to eat.

The fish feasted on the bugs, and it was not long before Matisha and her friends had grown to the size of *your little finger!*

But there were many dangers in these waters. Bigger fish and hungry birds saw the little fish as the perfect size for a tasty snack.

By not straying from the shallow water and using her lightning speed to dodge hungry kingfishers,

Matisha grew **and grew and grew!**

Before long, the not-so-little fish (*now the size of your hand!*) realized she looked different to her friends.

Some fish had colourful spots, others had pointed heads. Some fish were short and fat, while others were long and thin.

But Matisha was special!

Her entire body was covered in large, beautiful, golden scales. These shone like jewels, making her the envy of all her friends.

Soon, Matisha had grown too big to stay in the safety of the bay. It was time for her to find a new place to live.

With one brave whoosh of her tail, Matisha set off into the dark, swirling waters of the great Kaveri River.

The river flowed powerfully but
Matisha was a good swimmer. She
had great fun playing in the rapids
which created a fizzy underwater
storm of bubbles.

One moonlit night, Matisha found
a pool at the edge of the river
to rest.

As dark shadows silently moved
above her, the water's surface was
suddenly broken. An enormous
foot came crashing down, nearly
squashing her!

Matisha quickly moved to a safer
position and watched as a family
of elephants plunged their giant
trunks into the water to drink from
the river. After a while, they silently
moved back into the forest.

Matisha loved
exploring the
deep pools and
caves where
all sorts of
mysterious-
looking fish
lived.

One day, when she was watching some particularly odd-looking eels, there was a sudden BOOM! As the water FIZZED and FROTHED around her, Matisha lost all sense of balance and direction.

Matisha was terrified! She saw other fish floating past her. These poor fish were being scooped out of the water in nets. She was one of the lucky few to escape the local fishermen who had thrown explosive dynamite into the pool.

From that day on, whenever Matisha saw a human on the riverbank, she would quickly swim away in fear.

Over time, Matisha came to know that although most animals were friendly, they were also hungry. She had fun teasing the big crocodiles by darting away when they tried to snap at her. But she also had to learn to watch out for eagles! These giant birds dived into the river from great heights to try and catch a meal.

One day, she noticed that the water had a strange, sour taste. It was also dark and smelly, making it difficult for Matisha to see where she was going. Her tummy started to hurt and she found it hard to breathe. She also saw many dead fish floating past her.

Matisha made her way steadily upstream. The water became clear again and she was able to see that the black water was coming from a stream which joined the river next to a village.

At times Matisha felt lonely. All the other fish she had met had lots of friends that looked just like them, but Matisha had never met a fish that looked like she did.

One day, while she was wondering what sort of fish she was, big raindrops started to plip and plop into the river.

As the rain fell harder and faster, the water began to rise. The river got deeper and the flow got stronger.

The water smelled fresh and, without knowing why, Matisha started swimming against the flow. She swam through deep swirling pools and great rapids. By now the river was flowing so quickly that heavy boulders were rolling downstream. But Matisha
kept going.

As she powered her way upstream, Matisha noticed that she was not alone. Other fish of all sizes were dashing in the same direction. At one point, she was knocked aside by a gigantic fish. Quickly regaining her balance, Matisha saw a great big orange tail disappear into the water ahead.

Matisha did not know where she was going, but as the rain continued to fall, she continued her journey upstream.

Soon, the sound of the river changed to a deafening roar. Matisha had reached a deep pool beneath a great waterfall. A family of hungry otters were searching for their dinner and chased Matisha. They forced her to swim back downstream, where she found another deep pool to rest.

Just above this pool, a stream flowed into the mighty river. Matisha had never sensed such sweet-smelling water before and again started to swim upstream.

In no time at all, she found she could go no further. A huge wall of concrete was blocking her way. It also blocked the way of the other fish—fish which looked just like Matisha! She was overcome with joy.

During her travels, Matisha's tummy had grown. She was carrying a precious cargo of thousands of tiny eggs (*no bigger than a lentil each*). Along with her new friends, she carefully laid her eggs which sank into safe resting places between the stones at the bottom of the river.

A few days later, Matisha arrived exhausted at the great gorge from where she had started her adventure as a baby fish. As monkeys played on the cliffs above, she noticed tiny fish washing past her.

Content and happy, she
wondered if any of these
babies might be her own.

Matisha was now old and wise.
She had learned about many other animals
with whom she shared the river and
surrounding forests. She had learned how
humans can harm the river and the many
dangers that her own babies would face.
She now stayed in the safety of the deep
dark gorge and out of sight for the years
that followed.

One morning, as the sun rose and cast a warm glow across the great gorge, a tiger climbed down the cliff face to drink from the river. As he did so, something caught his eye—something he had never seen before.

In the water's dark depths a huge shadow lurked. The shadow drifted into the sunlight, revealing its enormous size, beautiful glistening scales of gold, and bright orange fins.

It was Matisha! She had grown into a majestic fish; one of the last giant hump-backed mahseer on the planet!

Matisha and the tiger looked at one another with mutual respect. Just as the mighty tiger was the king of the forest, Matisha was the mighty tiger of the river.

As the tiger retreated into the safety of the forest. Matisha, the 'river tiger', sank back to the safety of the deep dark pool.

Perhaps she is still there!

Why is the hump-backed mahseer important?

The giant hump-backed mahseer of South India's River Cauvery is one of the largest freshwater fish in the world and can grow to the size of a man. Unfortunately, this magnificent animal is now critically endangered and on the brink of extinction. Many other animals which live in and along rivers around the world are also in trouble. We all need water to drink and to grow the crops we eat but forget that we share our freshwater with many of the world's species. If we look after fish like Matisha, we are ensuring that our rivers remain healthy and able to provide for our essential needs.

Long live the tiger of the river!

The hump-backed mahseer is also known by its scientific name *Tor remadevii.* You can find out more about this amazing fish and what we can do to save our rivers here: www.mahseertrust.org.

Matisha is stuck in deep waters above a dam. She's ready to become a mother and needs to migrate upstream to a suitable area (marked A, B and C) in the dense forests and crystal springs to lay her eggs.

Some of these routes have recently been blocked by dams.
Can you help Matisha find her way before it is too late for
her precious eggs?

DR ADRIAN PINDER is a UK-based conservation scientist with a lifelong passion for the great diversity of fishes that swim in the world's rivers and oceans. Since his first trip to India in 2010, he has dedicated his research towards saving the iconic and highly threatened mahseer fishes of South and Southeast Asia from extinction. He strongly believes that children hold the key to securing the sustainability of Planet Earth's natural resources and aims to communicate his scientific knowledge to educate and inspire the next generation of environmental guardians.

When not exploring the remote jungles of Asia, Adrian can be found at home with his family on the Jurassic Coast of South England.

MAYA RAMASWAMY is a wildlife artist and illustrator based in Bangalore, India. As a schoolgirl, India's rich biodiversity drew her into volunteering for wildlife conservation. She is currently working on her latest collection of nature paintings, and continues channeling awareness towards Karnataka's vibrant natural heritage with children's books, hands-on workshops and live presentations.